INTI

From choosing his pa[illegible] to the name he gave to his encyclical, Pope Francis has been deeply inspired by St. Francis of Assisi. Indeed, says Pope Francis, St. Francis' love of God and of all creation is a spirituality we desperately need right now.

Our world home is in serious trouble, and in *Laudato Si'* Pope Francis adds his voice to all the voices worldwide calling for action. But truly effective action, he says, depends first on looking inward and renewing our souls in the spirit of St. Francis. Only then can we look outward to work for change.

Pope Francis is asking us to examine how our actions and attitudes are harming the Earth and the poor. And what about the world we will leave to our grandchildren and their grandchildren? What is our reason for being here? What will our legacy be?

Here are thirty days of reflections on this important encyclical. Each day begins with thoughts from Pope Francis. Let us take up his call to reflect together, pray together, and work together to renew our Earth.

The quotes from Pope Francis were chosen
by Deborah McCann, who also wrote the reflections
and the "Ponder" and "Pray" pieces.

Cover photo: Stefano Spaziani

a division of Bayard; One Montauk Avenue,
New London, CT 06320. 860-437-3012 or 800-321-0411,
www.23rdpublications.com.

ISBN 978-1-62785-125-1 ■ Printed in the U.S.A.

1 | WHAT WILL WE LEAVE BEHIND?

Our actions now show our children what our values are. Is this the legacy we wish to leave them?

The whole topic of climate change is so broad and complicated that it's all too easy to ignore it or hope that the "great minds" will come up with solutions. But Pope Francis is asking us to think much more concretely and personally: What world are we leaving our children's children, and what does this tell them about what we consider important values? What are we teaching them by our practices?

PONDER

Would I want my children and grandchildren to emulate my actions? Are my values good ones?

PRAY

God of creation, help me to see that your creation is fragile and beautiful. Give me eyes to see where I need to change how I treat your gift, and help me to do so!

2 | OPENING OUR EYES

God designed our planet as a place of beauty and fullness, without want. What have we done?

There is no question that technological advances, especially in the last two hundred years, have made our lives much more comfortable. At least those of us in the First World reap the benefits of these conveniences. But at what cost? Pope Francis cautions that the quest for new and better ways of providing more and more goods has relied on more and more of the natural resources that our planet needs to sustain us. What is our constant need for MORE costing us and our planetary home?

PONDER

How have the conveniences that I enjoy come at a cost to our Earth and the livelihood of others?

PRAY

God of creation, open my eyes and heart to ponder the Earth and its beauty, and help me to better care for it.

3 | WHO'S IN CHARGE?

We have lost sight of God's goodness by confusing stewardship with ownership.

In this encyclical, Pope Francis urges us to consider the human terms of what our consumerism costs. The constant need for more things and conveniences has put parts of the developing world in great peril—where those who rely on agriculture for their livelihood lose drinkable water and arable land, where sweatshops and factories replace the human dignity of achievement and pride in work. The world's imbalance of resources today shows how far we have fallen out of harmony with the divine plan.

PONDER

How can I follow the pope's lead by becoming more aware of how my choices affect others and the Earth itself?

PRAY

God of creation, forgive me for taking for granted many conveniences that those in other parts of the world long for.

4 | WE ARE NOT GOD

The land was given to us to till and nurture, not own and plunder.

One of Pope Francis' constant themes in *Laudato Si'* is our blind sense of entitlement at the cost of others' well-being. We were given dominion, not domination, over the Earth. By enslaving Earth and stripping it of its riches, we have left it breathless and in pain. Unlike the rich man who did not even notice the beggar Lazarus at his door, we have the chance to redeem ourselves and redress the wrong we have done to the planet. How do we begin?

PONDER

Do all the conveniences that surround me simplify my life or complicate it? Do they allow me time for contemplation?

PRAY

God of creation, help me to separate my wants from my needs, and help me to live more simply.

5 | THIS TIME, IT'S PERSONAL

Our care for the Earth is the same as our care for one another.

Pope Francis makes it clear that care for the Earth and its resources is inextricably tied to the people who are affected most by the way we are changing the Earth—the poor. All human beings are created in God's image—care for the planet, at its most personal level, begins with realizing our own interdependence.

PONDER

When was the last time I thought about who grows the food I eat and who makes the clothes I wear? Do they have enough to eat and a safe place to live?

PRAY

God of creation, help me to listen for the voices of the poor and marginalized, and give me the courage to speak for them and their needs.

6 | IT BEGINS WITH US

Before we can change THINGS,
we need to change US.

As a species, we have lost sight of our interdependence and reliance on one another, not just as producers and consumers of goods but as human beings alike in dignity and joined together with our Creator as one family of God. When we change our focus, how we treat our world will change. Pope Francis is very clear on this—by concentrating solely on the here and now, we have lost sight of our responsibility for the future.

PONDER

How much of what I buy, wear, eat, or read each day is produced locally? How much of my life relies on people I have never seen and will never meet?

PRAY

God of creation, help me remember that all people everywhere are your children—and my family.

7 | WE BELONG TO GOD

Our God-given gift of grace in our hearts will help us regain our human dignity...and restore the world!

No matter where we live, how menial our jobs, how rich or poor we are, we are all united in dignity as children of God. Pope Francis reminds us that this very gift is what will give us the power to improve our lot, to begin a new way of seeing and caring that will encompass the world as a whole. If we begin to treat each other with honor and respect, we can begin to address poverty in all its forms—personal, social, and material—and begin to treat the Earth as another brother or sister worthy of our notice.

PONDER

What do the poorest neighborhoods look like where I live? How am I connected to my brothers and sisters there?

PRAY

God of creation, help me to use your gift of grace to make life better for all, and to renew the Earth as our home.

8 | PRAYER IS A GOOD BEGINNING

Let us ask God for a good outcome to our work on behalf of the world.

Pope Francis reminds us that we have the greatest power on Earth on our side, the power of prayer, the power of seeking God's help as we try to take quick and positive action on the needs of all our brothers and sisters, including our Earth, so that the world we leave behind will still be viable for future generations. Prayer leads to action—and God's guidance leads to effective work.

PONDER

Pope Francis gets great inspiration from Francis of Assisi's *Canticle of the Sun*. What would my prayer of praise sound like?

PRAY

God of creation, guide my prayer so that what seems overwhelming may break down into something simple but effective that I can do.

9 | IN ALL THINGS, LOVE

Let us be guided by that most powerful force in the universe—love!

God created the heavens and the Earth to give us a home, and God entrusted to us its care and sustenance. Our present global crisis has been brought about by ignorance, greed, and the quest for power and dominance. Pope Francis reminds us that there is always one force that is greater than that—love: love for our Earth, love for one another, love for God, who has provided it for us. It is not too late for us to learn to express that love in how we care for all that we've been given!

PONDER

It's easy to look at my loved ones with love; how can I learn to see the enormity of God's creation with the same love?

PRAY

God of creation, when I stop to consider all you have made, it takes my breath away. Help me to take the time to praise you more often!

10 | BUT I'M ONLY ONE PERSON

If there is one person filled with hope,
that can make all the difference!

Pope Francis acknowledges that the problems facing our planet can seem overwhelming, and he lists many of them. Just one example: The destruction of the rain forests is killing off animal and plant species in horrifying numbers. But he also encourages us to see beyond the problems to the power we have to make changes. He reminds us that God has given us the grace and gift of hope to believe that our efforts will be fruitful. Let us use this gift well and not be afraid to let it shine!

PONDER

Do I look at the problems facing our planetary home as either distant and not involving me, or too big to contemplate? How can I change this view?

PRAY

God of creation, fill me with hope so that I may embrace your creation and work to heal it in whatever way I can.

11 | THE POWER OF GOODNESS

Never doubt that what you do influences change.

When we consider all the dimensions of the environmental crisis—from extinction to starvation, from change of habitats to loss of clean air and drinkable water, from wealth for the very few and endless poverty for most—our first reaction can be to shrink away, to think that we're simply not significant enough to cause effective change. Pope Francis says this is a false idea—that just as ripples move out from a stone thrown in a pond, so the power of goodness is recognized and imitated.

PONDER

Who has been a powerful example of good in my life? How might I try to imitate that person?

PRAY

God of creation, give me the courage to set my fear aside, and fill me with the confidence I need to help repair our Earth.

12 | IT'S AN INSIDE JOB

To make outward change, we must first look inward.

What will finally bring us to the realization that we must work for change? Pope Francis suggests that it's when we accept that we are intimately connected with all life on our planet, and that what hurts one of us wounds us all. When we have more to eat in a day than some do in a week, there's an imbalance. When factories pollute streams and destroy drinkable water, there's an imbalance. By embracing our interconnectedness—fish, birds, animals, and all our brothers and sisters—as God's creation, an immense love for our shared planet should begin to grow in us. Let's nurture that!

PONDER

How do the choices I make in a day affect the lives of my sisters and brothers globally?

PRAY

God of creation, help me to feel the sorrow of all of nature at its destruction, and the loss of species as the loss of family.

13 | WHO ARE WE, ANYWAY?

To know who we are, all creatures of God, will change our whole worldview.

Our culture tends to live as if everything is throwaway, resources are limitless, and all who say differently are uninformed and misguided. But Pope Francis reminds us that we are charged with a great task of stewardship of the Earth. We have to know who we are, brothers and sisters everywhere united as creatures of the one God who made us and our world. Knowing that, it is impossible NOT to work for justice and mercy—for our planet, for our neighbor, for ourselves.

PONDER

Who is my neighbor? Do I truly understand what stewardship of the Earth means in human terms right where I live?

PRAY

God of mercy, help me know who I am, and help me celebrate that identity with *all* my brothers and sisters.

14 | WE ARE ALL ONE

All living species are tied together in one great, interdependent network.

The smallest subatomic and atomic particles are joined together in a cosmic dance, and every living thing has evolved together on this world. There is no aspect of this amazing creation—chemical, biological, or material—that is *not* interconnected. Pope Francis says that our failure to note this in times past has led to today's problems. Our embrace now of how crucial every part is to every other part is what will lead to positive change.

PONDER

Where do I need to deepen my faith so that I may start to recognize our interconnectedness?

PRAY

God of creation, help me to see your glory in every part of creation, even where I'd least expect to find it. And if I see it being harmed or clouded, help me to reveal it to all.

15 | THINK GLOBALLY!

We need one plan for our one world.

When we start to think and act as brothers and sisters instead of distant acquaintances or masters and servants, our whole approach begins to change. We may begin with the most local of actions—recycling, using only Fair Trade products, biking or walking instead of driving—but as our consciousness grows, so does our commitment to global change. Pope Francis reminds us that, as we are all united as one family living on our common home, so we must have one plan for our world. The time to act is now!

PONDER

Is my parish "green"? What measures could we begin to take to model good stewardship of the Earth's resources? How can we start connecting our actions to what others are doing?

PRAY

God of creation, open my heart to neighbors in need. Help me build bridges between neighborhoods and families. Help me see you in all.

16 | ECOLOGY OF THE HEART

Our relationship with God and with one another informs our relationship with the environment.

Throughout *Laudato Si'*, Pope Francis returns to the contrast between dominion and domination, of stewardship and ownership. He tells us that our concern for the planet is an empty one if it does not take into account our responsibility and obligation to our poorest and least acknowledged neighbor. Our call is not to make things better for us, but for everyone. Since God is creator of all, God loves all. So must we.

PONDER

How can I become selfless enough to call everyone I meet a brother or sister?

PRAY

God of creation, it is not always easy to see you in others. When people are downright nasty and hard to help, give me a heart to see through their anger and fear to your grace shining in them.

17 | CONCERN FOR THE POOR

It's time for global initiatives, and this will affect how we treat the poor.

Virtually all climate scientists are making it plain that the time for drastic action on the environment is now, and they caution that it may already be too late to stop some of the damage. Some people reject the scientific consensus and say we need more time to study the problem. Pope Francis says that attempts to discredit calls for radical change come from the same forces that keep the world from addressing the issue of global poverty. Poverty has many faces—neglect of nature leads to neglect of humanity. He urges us not to continue our blindness but begin to reach out in love and compassion to the poor.

PONDER

Who and where are the poor where I live?

PRAY

God of creation, break the hardness of my heart so that I may hear the cries of the poor and then do more to work on their behalf.

18 | I AND THOU

We must be open to recognizing, accepting, and honoring the "thou" of each and every person.

In many languages there is a distinction between the "you" we call family members and the "you" we call acquaintances. Pope Francis reminds us that the more familiar "you," or "thou," is how God thinks of us and asks us to think of one another. What a change a word can make! When the other person is not an impersonal contact but someone whom we address as one of our own family, someone we treat the way God treats us, what responsibilities and opportunities open up for us all!

PONDER

How many "thous" are there in my life? How might I begin to widen my circle?

PRAY

God of creation, fill me with the wonder of your presence in everyone I meet so that we may all become one.

19 | THE WORLD BELONGS TO ALL

Now is not enough. What will we leave behind for future generations?

Pope Francis grows stern in his call for us to think about the generations that will come after us. Leaving a sustainable planet for our children, their children, and their children's children is not just a nice idea; it is a mandate of justice. This world is our gift to them. What will it look like? What are we leaving them to remember us by?

PONDER

What steps can I take now to reduce the impact I have on the Earth and its resources? What can I give back?

PRAY

God of creation, help me to see ways that I can repay the Earth for some of the bounty it has given me.

20 | WHO IS MY FAMILY?

Humanity holds a special place in this work for the world. We humans are not just statistics, we are creatures with an infinite dignity conferred on us by the Creator.

So much of the discussion of climate change and our responsibility to the planet involves dry statistics that are easy to ignore. Less easy is ignoring statistics with faces: the poor at our doorstep, the workers in farms, fields, and factories whose standard of living is low because ours is high. Pope Francis calls on us all to remember the human part of this equation—we are all important in God's eyes, and it is our responsibility to care for one another in new and more intentional ways.

PONDER

What does my local homeless shelter need for its guests? What can I provide?

PRAY

God of creation, when riches for one mean poverty for another, help me to seek the welfare of all.

21 | REMEMBERING HUMILITY

We need to remember that we are not the "masters of the universe"—God is.

Pope Francis makes it clear in *Laudato Si'* that we have allowed our own individualism to overcome our need for God. We have treated the Earth and its treasures as put here for our own convenience rather than as gifts to be shared by all. Is this the kind of value system we want to pass along to future generations? Pope Francis' answer to this is a ringing "no," and he reminds us that making God the center of our lives is the best way to begin to acknowledge and accept all our brothers and sisters.

PONDER

How central is God in my life? How can I improve our relationship?

PRAY

God of creation, it is so easy for me to think I'm in charge, that one small failing won't affect much. Help me to realize that everything I do has an impact for good or ill.

22 | REACHING OUT

All creatures are connected and all are worthy of love and respect.

Pope Francis stresses again and again in *Laudato Si'* that our constant push for more and more control over nature has led not only to terrible harm to the natural world but to the dehumanization of us all—those in control lose their humanity spiritually, while many of the people who are controlled lose their homes, their livelihood, their own human dignity. We have the power to reverse this, if we begin to see ourselves as the interconnected creatures we are, learning to cooperate and live together as family.

PONDER

Can I try today to look closely at someone I might otherwise ignore and to see them as family?

PRAY

God of creation, help me not to want more and more but to appreciate the many gifts you have already given me. Help me to share these gifts with others.

23 | NO ROOM FOR INDIFFERENCE

The time has passed when we can stand alone.
We must stand united...or not at all.

When astronauts sent back their photos of our tiny globe, for a moment nations seemed to grasp that we are all one. In this time when we are in danger of destroying the only home we have, Pope Francis urges us to forget borders and barriers, put aside our autonomy and possessiveness, and begin to embrace one another as the one family we are on the common home that is our Earth.

PONDER

Of all the things that I consider "mine," are there any that really deserve that title? What would I not share with another?

PRAY

God of creation, we love our fences, our doors, and our locks. Help us to remember all those persons without the privilege of security and privacy, and help us to live more communally.

24 | MAKING OUR COMMITMENT SINCERE

Care for nature without care for one another is empty. We are called to love one another.

Some people talk about "nature" as if humans don't matter at all. Pope Francis reminds us that the entire web of connectedness embraces all of creation, and that caring for nature means caring for humanity and realizing every person's essential human dignity. Do we have to look more closely to discover this sometimes? Of course! And that's exactly what will begin to bring about the change needed in society. Recognizing that we are alike, not different, will help us survive.

PONDER

Am I truly present to all the interactions I have in a day?

PRAY

God of creation, take away my fear of people and things that are different so that I may see our common bond.

25 | HEADING HOME TOGETHER

We are all joined in the great enterprise of bringing one another to God.

Pope Francis reminds us that our purpose in showing love to all is to model Christ, whose whole life was devoted to bringing us to the Father. When we imitate his example by honoring one another, our goal will be to ensure a healthy planet for all. All are deserving of decent places to live, the right to meaningful labor, the right to peaceful coexistence in a world that supports and celebrates that life. All thoughts of tyranny and domination pale in the reflection of this divine mandate.

PONDER

Is there one thing I can do today to make my neighborhood a better place?

PRAY

God of creation, help me to remember my part in your plan. Help me to follow the example of your Son in all that I do so that I may reflect your light to others.

26 | LOVE WILL GUIDE US

We are called to build a "culture of care" that will help us renew the Earth.

When we realize our own part in creation—how all-encompassing and intertwined care for one another and care for the planet are—we begin to realize all the implications of what God is asking of us. Pope Francis says that it's time to leave self-interest behind and begin to embrace a "culture of care" that will open us to the needs of all our brothers and sisters and lead us to protect our planetary home as we protect one another.

PONDER

What would embracing a "culture of care" mean for my life? When I'm faced with the tasks before me, do I remember to ask for God's help?

PRAY

God of creation, help me to see your hand in everything, so that everything I do in your name, I may do with love.

27 | FROM INSIDE TO OUTSIDE

God is everywhere and in everyone and in everything. God fills the universe.

One of the first things we learn as Catholic Christians is that God is everywhere and in everything. Yet our actions—subduing habitats, polluting the Earth, and considering some people or places as disposable—indicate that we have either forgotten or ignored this central tenet. Pope Francis reminds us that God resides both in the soul and in all things. Let's begin to celebrate God's presence in everything and everyone around us. God is not our personal possession; we are meant to open ourselves to God's actions through us.

PONDER

Am I truly aware of God's presence in my life?

PRAY

God of creation, open my heart to let you out, so that I may work with conviction to bring about a better world.

28 | LOVE IS THE ANSWER

Love is the force that will guide us to bring about needed change.

In *Laudato Si'*, Pope Francis speaks warmly of God's love, a love that is pure and unconditional, and he asks us to believe in its power to transform us. This is the force that will guide us and open our eyes to the needs of our brothers and sisters everywhere. Caring for them and their needs, in the love we share as children of God, cannot help but lead to rebuilding the planet so that all may rest safe in God's creation. God promises to be with us in this work. What have we to fear?

PONDER

Do I really understand the power of unconditional love? Have I known it in my life?

PRAY

God of creation, fill me with your love so that I may joyfully and lavishly share it with others—even those I have never met, for we are all part of your family, and you love us all.

29 | IT'S IN OUR HANDS

We should not despair. With God's love to guide us, we can replace irresponsibility with attention, and problem-making with problem-solving.

Pope Francis says bluntly that our post-industrial society may well be looked upon as the most irresponsible of all time. But he also holds out great hope—rooted firmly in Scripture and guided by the example of Christ—that we have the ability to reverse course and become known as the generation that did make a stand for restoring the Earth and building a society founded on co-responsibility, co-operation, and mutual respect and understanding.

PONDER

Can I overcome my own fear and indifference to work for change? How can I get started? Do I believe in God's belief in me?

PRAY

God of creation, help me to adopt St. Francis of Assisi's lavish and rapturous praise of you and the world you have made!

30 | GO MAKE A DIFFERENCE!

May we be converted by God's limitless grace to go and change the world!

Pope Francis does not deny that we are faced with very serious challenges if we are to leave a sustainable world for future generations. Nonetheless, he exhorts us to be joyful in our confidence that God's love and grace will help us in the quest to replace poverty of resources and poverty of soul with the richness of shared unity and purpose. He asks us to live out this challenge as a true conversion of heart and soul. Let us all stand united in wonder and awe at the God who made us. *Laudato si'*—Praised be!

PONDER

Am I ready to face this challenge? How shall I start? With whom can I join forces to effect change? Can I accept the joy that God wants to share with me?

PRAY

God of creation, help me share your light in all that I do to renew and restore your world. Fill my heart with action and praise!